The Cocker and The Game Cock
Information for Breeders and Amateurs of that Noble Bird, the Game Fowl

by W. Sketchley

with an introduction by Jackson Chambers

This work contains material that was originally published in 1814.

This publication is within the Public Domain.

*This edition is reprinted for educational purposes
and in accordance with all applicable Federal Laws.*

Introduction Copyright 2018 by Jackson Chambers

Self Reliance Books

Get more historic titles on animal and stock breeding, gardening and old fashioned skills by visiting us at:

http://selfreliancebooks.blogspot.com/

Introduction

I am pleased to present this third title in the "Game Fowl" series.

The work is in the Public Domain and is re-printed here in accordance with Federal Laws.

Though this work is a century old it contains much information on poultry that is still pertinent today.

As with all reprinted books of this age that are intended to perfectly reproduce the original edition, considerable pains and effort had to be undertaken to correct fading and sometimes outright damage to existing proofs of this title. At times, this task is quite monumental, requiring an almost total "rebuilding" of some pages from digital proofs of multiple copies. Despite this, imperfections still sometimes exist in the final proof and may detract from the visual appearance of the text.

I hope you enjoy reading this book as much as I enjoyed making it available to readers again.

Jackson Chambers

Disclaimer

This book was written in an age when cock-fighting was widely acceptable throughout society. In many places throughout the world, cock-fighting has been made illegal.

The material presented herein is intended to be strictly for educational purposes with the purpose of enlightening Game Fowl breeders about the history of their breed. Publication of the material is neither an endorsement, nor a criticism of its contents. This book is presented as part of large series of educational material on the history and raising of numerous chicken breeds for utility or exhibition purposes.

As the reader, please consider it your duty to become familiar with local, state, provincial and federal laws relating to the subject matter contained herein before attempting to utilize any of the information presented.

As the author, publisher and retailer cannot control how the reader utilizes the historical information presented in the pages herein, they hereby disclaim any liability to any party for any loss, damage, disruption or other liability that may be incurred by the reader's misuse of this material.

PREFACE.

The author having been attached to the sod at a very early period of life—and having lived in a part of the country high in repute for that noble bird, the Game Cock; where as great a variety of birds were exhibited in all their various process of refinement as any individual amateur could enjoy, he flatters himself that by such superior means of collecting information his attempt at writing "The Cocker" will be found to contain not only instruction, but be a source of amusement to the reader, and as such may serve to plead his apology for undertaking the present task.

Under such a pre-eminent latitude of superior breed, no man of discernment could be at a loss (to mark those whose superiority of blood, mode of fighting, with all the necessary concomitants that constitute perfection in the game cock) how much to appreciate their worth.

To collect a stud from these renowned war-

PREFACE.

riors was the pleasing task of the author, whose situation in life enabled him to obtain the object of his pursuit. But our juvenile endeavours are not always adequate to the difficulties attendant on true breeding; for notwithstanding the judicious choice of cocks so made, we are liable to fail from some incompatible choice of our ideas. To unite their crosses, (in order to complete a uniform set of birds,) similarity of feather, constitution, colour of the beak and legs, are in themselves so requisite, that for the want of a monitor in our early progress we too precipitately anticipate all we want in procuring these models of perfection, when, in fact, we have perhaps accomplished very little essential to the constituent part of breeding. The author trusts that an extensive routine of practice for fifty years, aided by the acute observations of others, and committing to paper from time to time those of his own, has enabled him to afford such information as may pilot the juvenile breeder through so pleasing, yet so arduous an undertaking; knowing well that observations founded upon experience, are the surest guide to truth in every science, and are more likely to succeed than the most refined and plausible theories.

March 30*th*, 1814.

THE COCKER.

To Breeders.

In our early attachment to, and pursuit after, cocking, and on our commencement of breeding, we are frequently hurried on in our choice of a single cock from a day's fight, or a promiscuous match, that has exhibited appearances of courage and heel. Although condition will not give heel, yet it very much promotes courage, inasmuch as to carry an indifferently-bred cock through a tolerably well contested battle.

From these superficial trials and our own want of experience we are induced to purchase cocks of this description to breed from, which must inevitably involve us in error. But maturer experience will inform us, that high condition will incite courage in a half-bred cock; and a bloody heeler in that situation may succeed against one of superior game and less heel. It therefore shews how necessary it is to guard against a prepossession of any single cock however meritorious.

To avoid this youthful partiality, if you attend regular mains, you will have an opportunity of seeing a number of well-bred brothers, whose mode of fighting and supporting their battles in any stage, mark them for steady fighters, good heelers, ready mouths, and deep game. Such proofs would warrant your endeavours to procure one of these warriors the least injured in his contest, more immediately if their feathers and other concomitants are to your wish. Practice and acute observation will tend to inform you what feathered birds are the most approved, as well as the best in constitution. But of those endowed with every sought-for requisite, the black-reds are most in estimation. The difficulties attendant on our early endeavours are only to be surmounted by a steady attention to those particulars that are requisite in this pursuit; and when you have fixed on such cocks, whose originality is well known, and whose breed has stood the test of experience, it will not be difficult to procure such, either by interest or connexion. If hens or pullets cannot be obtained (which should if practicable) you must be equally cautious in selecting them from other sorts similar in feather, constitution, and every other attendant. Having once arranged your brood, conform to your directions: and at a proper season you may commence

Breeding.

GENTLEMEN who have been in the habit of breeding, may have, from keeping the old brood too long together, brought their cocks to a stand still; when by a judicious mode of keeping in and in, that circumstance might not have taken place. It is generally admitted, that a cock is in full prime and vigour at two years old. But how frequently we find that cocks are continued with hens until they are six or seven years old. At three years old it is well known that he begins to lose that sprightly bloom he wears at two; his length of plume encreases, and his hack exhibits too loose a texture, dangling over his throat. When this is discerned, we ought to substitute another in his stead; for he will become inactive, languid, and balk his craving partners—and you may breed in vain. The hens may in general be held to a longer date, as they retain the power to propagate beyond the period allotted to the cock. If this opinion is granted, the impropriety of continuing them beyond that period may account for the slow and inactive cocks so frequently met with.

To elucidate the foregoing remarks, in order to enable you to judge of the constitution of the brood-cock you mean to select, he should have every apparent feature of health: such as a ruddy complexion—his feathers close and short—not cold or dry—flesh firm and compact—full breasted, yet taper and thin behind—full in the girth—well coupled—lofty and spiring—a good thigh—the beam of his leg very strong—a quick large eye—strong beak, crooked and big at setting on—not more than two years old, put to early pullets, or a blooming stag with two-year-old hens—and when a cock, with pullets of his own getting. Great experience justifies me in vouching for the prosperity of the practice. In order that you may attain a sufficient number of in and in for your establishment hereafter, March, April, May, and June, if they have been early together, are the months to propagate: the latter month will not be too late for two-year-old cocks. For the two first days of a long main very early chickens have their inconveniences; and if you have none at hand till April they will probably be esteemed sufficiently early, as the springs are not so congenial for breeding as formerly.

In the choice of your hens let them be rightly plumed to your cock: nor let your choice fall upon those that are large but rather suffer the cock to make up deficiency in the hens being small: their shape should be similar to the cock—lofty necks—short and close feathered. A true blood hen is seldom or never gummy in the bone of her leg, but clean, sinewy, and, in length, proportionate to the rest of the body, with a well-set thigh, long, clean, and taper toes, so that they may, as far as is practicable, be as near in every respect to your original brood, as the nature of breeding will admit of. Deviation will take place in feather, &c. instances of which have frequently occurred with me, where Dame Nature has interfered in her varying change: that notwithstanding every well adapted system, she will prevail in surprizing us with some productive alteration which inexperience is at a loss to account for. Fifteen years or more I had enjoyed an invariable production of the most complete black-reds bred by any amateur, without a single instance of deviation during that period, but on the sixteenth year I had several light Piles in one hatch;—no change of eggs could possibly take place—or was there a shadow of doubt of interference with any other cock, but a strong recurrence to the Pile at that distant period. A well regulated account of my cocks enabled me to ascertain that there had been a Pile in the cross five years previous to my having them out of Shropshire: so that they held highly regular for twenty-one years, not only in plumage but in every desired requisite. To have gone on with this deviation would in the end have produced Spangles, &c. Of course the pullets were cast aside, and the cocks fought off. By a persevering conduct in selecting the darkest, or those most resembling your original attachment, is the surest criterion you can possibly pursue in keeping up, not only their feather, but constitution.

Let then no recommendation, however high, induce you to breed from cock or hen (when you find it necessary to make a cross) either with one or the other, that differs essentially from your old brood; for notwithstanding every judicious caution in the selection of that choice, the introduction of a new cross (though every way similar) may prove more injurious than the evil you are endeavouring to correct; therefore in this case your enquiries and ocular demonstration should be your peculiar care.

Your brood now selected under every due consideration, I should recommend from four to six pullets or hens, being as great a number as should be put to a cock, (probably only four hens to a stag) and should always prefer a maiden cock to a fought one, of the same sort; the former suffers no inconvenience in being fed and tendered by hot meals, close covering, clipping, ruffled temper, loss of blood, and receiving unknown injuries,—whereby the intention of breeding may be materially frustrated. On the contrary, they are endowed with nature's best gifts, and of course best calculated to answer the desired end. I should advise their being put together as early as November or December, that the cock may be perfectly at ease on his walk, and have ample time for ingratiating himself with his new acquaintances. Every gentleman has an opportunity of selecting proper places for breeding, where they can experience as little interruption by dogs and vermin, as possible. The distance from any other house where fowls are kept should be at least half a mile—the situation, a dry gravelly soil; and it would be better if they could enjoy a constant spring of clear water, and shade, the nearer your brood-place the better. A situation where cocks are liable to interfere with your hens has too frequently occasioned the failure of supposed brothers in the same hatch, and has been the cause of greater derangements in the course of breeding than any other circumstance whatever. Another necessary caution is, not to suffer any hens different in breed to be turned down with any set of sisters; for notwithstanding every caution in selecting eggs, you may be deceived, or held dubious, and ruin every well projected plan. Have a marked attention to see that your brood cock bears himself well to all his hens. It frequently happens that one or more labours under his displeasure, an antipathy we are yet at a loss to account for; in that case they become useless and should be removed. If his gene-

conduct be severe, I should suspect him of cowardice, for several cocks of this description in the course of my breeding have turned out but indifferent. You will find that those hens under this plumed bashaw's arrogant dislike are generally held at an awful distance, and are seldom or never attendant upon him, but are recluse and solitary. This may account for the many unprolific eggs in extensive breeding. Whenever you have occasion to remove any, be the cause what it will, do not disturb the repose of the rest by turning down a fresh hen; for by such a change I have been deprived of the use of the whole for that season, nor should it be practised at any brood department or elsewhere where valuable hens are, without observing their conduct towards each other. Such has been the conflict upon those occasions, that they have never recovered their usual gaiety and constitution, but gradually pined away.

The saving of eggs too early, which you intend to set, from hens that have not been with your brood cock from the beginning, has occasioned such errors as time alone can correct; and although even a few may only partake of the prolific stamp of another cock, those few may be productive of incalculable mischief, and engraft a blood so every way different from your own (the cause too frequently observed) that in the course of fighting, you have good and bad cocks supposed to be equally bred alike, and you are embarrassed to know how to remedy the evil. This circumstance and another equally if not more dangerous (the unwarrantable practice of changing the eggs) every Fancier who has been in the habit of long breeding, has experienced. To obviate such errors, so as not to admit of a doubt (at least as far as human foresight can insure us) have them so early together as may in that respect totally exclude the idea:—even with this precaution, I would not save the few first eggs, nor the last.

It is not in the compass of practice to avoid every particular inconvenience that arises in breeding, or I could wish that every hen could lay distinctly—that the eggs might be marked differently, and of course hatched under separate hens: for every sister may not enjoy an equal share of good health. In this particular, how requisite it is, that the person employed or engaged in the pursuit should have a knowledge of those deviations of health to which these birds are subject in order that they may be detected as early as possible. It does not always fall to the lot of those whose province it is to superintend them, to discriminate to that extent: but the many inconveniences and errors attendant on this pursuit, should point out the necessity of their being possessed of such qualfications. A single day should not pass without seeing the general brood, and the utmost attention paid to every minutiæ to prevent the errors committed by neglect; and more particularly so, when we experience some default, notwithstanding our utmost care, for it certainly behoves us to exert our best endeavours to bring them into the hands of the feeder, in whom we confide, as unexceptionable as possible.

'Tis doubtless an absurd opinion to think any breed incestuous that springs from the brute creation, and of course we have bred from father and daughter, mother and son, or from brother and sister, which is termed full blood. I have also known the brood excellent where the brood-cock and hens are got by the same cock, but out of a different hen. Though I most approve of the former, the hen's strain being generally allowed to be superior and more certain than the cock's. If your brood places are at a distance from your house or place where you mean your old game hens to sit, great care should be taken that your eggs in being conveyed away, are not cracked or shaked, but compact and firm for carriage. As eggs are best marked when gathered, always on these occasions provide yourself with pen and red ink, and mark each egg with some character known to yourself, with the day of the month: for, as you may not always have broody hens ready, this method will point out to you to set or destroy them according to the time they may have been on hand. I have generally kept mine in sweet bran: their own weight imbeds them and prevents their contact: cause them to be turned every two days—for by lying too long in one position, the yolks will frequently decay, and destroy the prolific power. It frequently happens that some eggs are smaller than others, and ill-formed—they therefore should be rejected; for all misshaped eggs will produce defective birds. From intense observation I have generally found that the round egg produced the female, and those of the oblong the male bird. If too many eggs are set under the hen so as to be exposed to the chilling cold or too intense heat, either extreme impairs the vital power, and the embryo will prove deficient. Nature prompts these creatures to turn their eggs during incubation; equally necessary is its being done previous to their being set.

A few years will provide you with a sufficient number of old game hens to sit, and on no account be prevailed on to use any other. Old hens are always more steady in sitting than pullets, are more industrious and attached to their brood, and not half so prone to quit their brood at too early a period. Their places for sitting should be private, free from annoyance, and ought to be as little ruffled as possible, save more immediately to see that they are not laid to, as well as to observe that she has not deserted them :—to give them every chance of secure retirement, they should be little liable to intrusion. It has been recommended to supply them with food, &c. near them. Whatever is most natural I should think most conducive to their health, and therefore have suffered them to come off to enjoy good water with feed at a certain place, that they may not be too long absent from their eggs, with any other enjoyment they are in search after.

It is a good and regular method to chalk over the place where they sit the day they should hatch, and of course draw your attention to see that the eggs are perfectly right as to number, mark, &c.—and to remove the chipped envelope, as well as the chickens which are hatched, until the whole are at hand. You will then return them to the hen in such place as you may have for the purpose (boarded floors are best) where they remain as to time according to the clemency of the season, and the strength of the chickens. Let their feed be

> Macerated eggs that have been boiled hard;
> Crumbs of white bread;
> Lettuce leaves, well mixed with an addition of meadow ants;
> The maggots from grains, kept for the purpose;
> Shilled steeped oats;
> Small wheat;
> Curds, with new milk;
> Bread toasted, steeped in chamber-lie,

as they are fond of variety. Let their food be given frequently, in small quantities, and accommodate them with small heaps of dry earth or fine sand in the room.

You will observe never to carry them abroad until the dew is entirely off the grass, every kind of humidity being hurtful; and you will return them before sun-set.

As more hatches than one may be in the same place, never delay marking them when brought into the room with some one of the marks usually put upon them (perhaps those upon the nostril and eye are the most injurious) in order to discriminate their sorts as well as to enter them in the manner set forth in the book for the purpose, previous to your leaving the room. If you have plenty of range in your department, a great many chickens may be kept until such time as the hens may leave them. A distribution of your cockerils claims your serious attention, so much so that one half of your early birds may be preserved until such time as they are to occupy their allotted walks. The mode I have pursued to accomplish this desired end is, to select as many early cockerils, as nearly of an age as they present themselves, and turn them down in some secure retreat, under the guidance and authority of a two-year-old cock, with one hen. Here they may remain without a probability of their becoming rebellious or self-contentious until November, when it may be right to do away the old cock and his mate, and suffer them to enjoy an uncontrouled retirement until a proper disposal offers. No other mode that I could possibly devise has offered me so much security as this, and what I should strongly recommend. Others you may dispose of in such farm-yards, where interest is most predominant, and there remain a proper season until their removal. Frequent visits are in this department necessary to watch their growth and well-doing—and at a proper time to make choice of those whose shape and perfection promise to reward your future care. Those that are not the objects of your choice, see them properly disposed of, in order to prevent an improper use.

Previous to their going to master-walks have them up for twelve or fourteen days, that you may cut their comb and wattles; and handle them with gentleness and every encouraging demeanour. Let them go proudly out of hand, and touch them lightly behind to bring them to the front of the pen: this will feed their pride, inure them to the crow of others, and they come to with more alacrity and pleasantness to be fed and with more facility than cocks unaccustomed thereto; and from this necessary attention few or any are liable to shy.

Let your pens be well aired, the fastnesses properly secured, the perches arranged, the straw sweet and not damp, and every morning shook from their filth. Before you send them out number your pens from number one to the number you have up, with the person's name they are to go to; and having your book ready enter them as directed, being particular as to their marks and colours, with any other natural mark they may have, and ticket your bags according to the pens, when sent. A regularity of this kind will save much trouble.

Your utmost care and attention must be exerted to procure good walks, for half-bred fowls in a well furnished walk will beat the best game when starved or pined; and hand-strewed walks generally bring on an inactive sloth. To send fine stags that have enjoyed every indulgence to bad walks, is one of the most flagrant errors a breeder can commit, and it is undoing all you have done before. Cocks, from so sudden a deviation, experience a change in their system, and it checks their growth—frequently a gradual decline ensues. Therefore the procuring good walks is absolutely necessary and conducive to the well-doing and constitution of your cocks. All town-walks, except here and there a few, are not worth having, and there are few in villages where towns are near to each other, but may be ranked in the same class. The best are those whose situations are distant, and where plenty of corn and water abound. Grass walks with corn are to be preferred to clay-bound fields, the latter defacing their glossy plumes. Where a great number of walks are wanted, the practice of running stags with cocks is unavoidable, and with some to a late period; even if he fights a long main early in the spring he may fall short of the whole of his stags being got out, and of course a many are sacrificed. If you have much yard-room, or two yards belonging the same dwelling, let the younger brood be accustomed to occupy the one, with a proper roost distinct from the other, seldom interfering with the older branch. Gentlemen who command any number of walks, have infinitely the advantage of those whose walks are few and limited: the advantages over the latter are pre-eminently great, for many are so beautifully situated that even the crow or the sight of a cock seldom come across them; they are neither fretted nor teazed, which ever causes them to lose much of their flesh, and destroys that martial fire and spirit, when so habituated, added to the annoyance of stags—that when exhibited upon the Pit, his raging pride is so far abated, it frequently makes him tardy and slow to action.

Those who fight for considerable sums cannot be too scrutinizing in the choice of their stags, when they are to be sent out to clear walks, to see that they are in all respects free from ocular imperfections; for the occupying walks with any deficit is not only an increase of expense, but a great disappointment, as it frequently happens for want of such nice observation, that they are reckoning upon more fine cocks than they are possessed of.

To mention a few of these imperfections may be necessary, although they are generally well known; such as are

> Flat sided, and then generally deep keeled,
> Short legged,
> Thin thighs,
> Crooked or indented breast,
> Short thin neck,
> Imperfect eye,
> Duck and short footed, and
> Unhealthful,

may be easily seen when up for the purpose of cutting and handling.

Cocks that are well formed and lofty have an amazing advantage over the disproportioned; the latter carrying with them much useless weight. High bearing fowls will always have the odds in their favour over low setting cocks. Cocks when they are justly formed, rise in their fight with more agility and force, are better heelers than those that carry their make equal to the extreme; and your dry heeled cocks are generally of the latter description, the weight being too far from the centre of action, and once overpowered they are always under a cock, that is not alike defective;—their legs are thrown out of the line of the body, and of course are never close hitters.

THE COCKER. 15

Cocks that do not bear cone-like shapes, are for the most part wide and straddling in their walk, and as they walk they fly—whereas in the cone-like shape, the legs are more inverted and narrow, and are more terrible in their spur.

Trials.

TRIALS are absolutely necessary:—Cocks vary so much in constitution, from one period to another, that how far a fair trial can be had from such as are imperfect is too well known for me to comment upon. I shall therefore only here observe that a trial of stags is very indeterminate; for they may be excellent in stags and very indifferent in cocks.

The variety of cocks bred in this kingdom, and the opinions of men being as various, it is difficult to say what sort to recommend in preference to another; for in one part of the kingdom they are partial to Piles, in others to

>Black Reds,
>Silver black-breasted Ducks,
>Birchin Ducks,
>Dark Greys,
>Mealy Greys,
>Blacks,
>Spangles.
>Furnaces,
>Pole Cats,
>Cuckoos,
>Gingers,
>Red Duns,
>Duns,
>Smoky Duns,

in all of which good birds may be found.

It has always been a matter of surprise to me, to see the wonderful avidity, even in experienced breeders, in expressing a wish of obtaining a single cock from a day's fight, that has exhibited something out of the common routine of play, in order to breed from, when I have been sensible of the impro-

THE COCKER.

priety of the cross he was destined to make, in fact, with hens that were as dissimilar in feather and other necessary similarities, as possible. If uniformity in their general appearance is absolutely necessary in forming a regular breed, I cannot help expressing my wonder at well-informed men running into an error so fatal to the welfare of judicious breeding; and which must convince a reflecting mind, that from such unnatural or at least incompatible crosses we are indebted to the public for such a strange medley of colours as we see in every main, when a few years' attention would exhibit cocks of a very different stamp. From such incongruous mixtures, we see

1.

The Pheasant-breasted Red,

2.

The large spot-breasted Red,

3.

The blotched-breasted Red,

which are all produced by some inaccuracy of breeding.

4.

The Turkey-breasted Grey,

5.

The large marble-breasted Grey,

6.

The large spot-breasted Grey,

have a cross that does not belong to the true Grey.

7.

The shady-breasted Birchin Duck,

8.

The streaky-breasted Birchin Duck,

9.

The marbled-breasted Birchin Duck,

have also a cross different from the true feather of the Birchin Ducks.

When I assert that No. 1 is from some inaccuracy of breeding, it must be understood that they deviate from the character of

The true black-breasted Red,

by the introduction of a cross of the Pheasant-feathered Cock, or a variety of the Spangle, either of which must deteriorate the original: and that No. 2, under

The large spot-breasted Red,

has been introduced either from the cock or hen in some distant cross of a black, and of course attaches some remains of that colour either in spots, streaks, shades, or blotches, which strain operates as an injury to

The true Black-breasted Red.

The remarks upon No. 2 hold good for No. 3.

The Turkey-breasted Greys under No. 4, 5, and 6, are under the same injudicious distant cross of No. 2, and wherever the distribution takes its seat (except a regular tip of the wing) they are inconsistent with that of

The true Mealy Grey.

The shades, streaks, spots, blotches, and marbled, whenever they differ in colour, their varieties arise from a cross possessing those colours at some distant period.

A regular and well chosen system to breed uniformly not

only in feather but in each character respectively is the best mark or criterion of an experienced breeder. When a main exhibits a regular set of brothers that require minute discrimination to distinguish one from the other, it meets with the general plaudit of the surrounding pit.

The feather of the True Black-Breasted Reds should be a clear vivid dark red, without any shade of the black whatever, extending from the hack to the extremities—the red upon the hack above, and black beneath, the upper convex side of the wing equally red and black, even those surrounding the posterior—the whole of the tail feathers black, the tip of the wing also—with black beak and black legs.

The brood-hen for such a cock should be the Dark Partridge-coloured Hen, bright red heckled above, black beneath, clean brick breasted, and such to the posterior; black beak and legs.

The Mealy Grey,

which may be ranked next in value to

The true Dark Grey,

is originated from the Black and Mealy White, has been the produce selected from those whose feathers were nearest to the Mealy White, slightly tinged and shaded with black; they have been kept in and in, and established the Mealy Grey, and from those of darker varieties have nearly all our Greys originated: the hen's colour will wonderfully prevail, in general more so than the cock's.

Number 7, 8, and 9, as remarked, differing from

The true Birchin black-breasted Duck-Wing,

are from an introduction of some broken feathers either in the cock or hen, and will gradually infuse those different traits.

The true Birchin black-breasted Duck

has been originally bred from the Black-breasted Red, the Yellow Birchin, and the Grey Duck-Wing Hens.

The feather of No. 3, (Birchin Duck) is a grey heckle tinged with black above, and black beneath, yellow ground with a general shade of dark Birchin, thorough and clean black breasted, yellow legs and beak.

No cocks exhibit a longer period of unfaded health than the True black-breasted Birchin Ducks; and their reputation stands as high in the opinion of sporting men as any general established feather going. When we come to appreciate the cross, you have in them as many excellencies in regard to feather, heel, courage, constitution, and shape, as will warrant an eulogium from the most experienced amateur.

The Piles have originated from a variety of crosses, and which have constituted the many shades you find in this numerous class. There is a strain in these cocks that eminently distinguishes them in that most wished-for gift and excellence—the deadly heel—that generally stamps their prowess in fighting; and it may be here remarked that the lighter colours wield their well-tempered weapons in a more dangerous direction than any other class of cocks.

Their admired excellence is that of close hitters. The true Pile was probably from the Ginger and light Custard Hen, and then crossed with Reds and Red Duns.

We have justly to regret that we cannot enjoy all the requisites in the Piles,

    Gingers,
    Mealeys,
    Yellow Greys,
    Blacks,
    Pole Cats,
    Cuckoos,
    Furnaces,
    Spangles,
    Custards, and
    Cuckoos.

THE COCKER.

They are so liable to intermediate changes of the constitution, become degenerated, soft, and long feathered, delicate in habit, and you find it an arduous undertaking to keep them for any length of time, to any fixed or established excellence.

Breeders of the present day are avoiding those injudicious crosses, and by adopting more uniformity are doing away the many party-coloured birds, which marked a certain want of refinement in breeding, so every way necessary to establish a choice few, which might be restricted to the undermentioned, and would in the course of a little time totally obliterate all others :—say

No. 1. Dark Reds,
    2. Dark black-breasted Red,
    3. Dark black-breasted Birchin Ducks,
    4. Dark black-breasted Berry Birchin,
    5. Silver black-breasted Duckwing Grey,
    6. Clean Mealy Grey,
    7. Dark black-breasted Grey, and
    8. Red Duns.

The whole of these if bred with care may be produced to a standard of uniform regularity, taken in their progressive numbers : thus

No. 1. Crosses well with the Partridge Hen, with No. 2 and 3 Hens.
    2. Crosses well with No. 2 and 3 Hens.
    3. & 4. Cross — As a cross already they are best kept to their respective feathers, always in youth.
    5. 6. 7. Will cross with each other with his own feather, invariably.
    8. Will cross with No. 1 and 2.

Many of my readers no doubt will say, that a good cock of any colour is acceptable, and individually may be esteemed so; but a regular pen of cocks uniform in feather, blood, and constitution, must have the advantage of an opposite adversary who claims his support from the many.

Piles are not in the number of my selects, although many that I have seen fight have had a share of my admiration, from a peculiar mode of fighting, singularly their own, their appearance being attractive and prepossessing. But all my endeavours to breed them have ever been unsuccessful; and as what I have to offer as a guide to others is from the sober light of experience, rather than from the delusive glare of plausible theories, I have omitted them in my favourite select. Perhaps it is too late in the day for me to make any further attempt to establish the true Pile, yet blessed at this time with a due share of undisturbed good health, I purpose to make an immediate trial. An undeniable cock has been offered that fought at Manchester, whose character admirably points him out as a proper object—and am now in anxious search for a suitable mate. My success, if I live to accomplish it, will be better and more ably detailed by a gentleman who may be induced to publish a second edition of this work.

REMARKS

ON THE

Change in Constitution.

It is one of the most essential requisites that appertains to the welfare of producing an unexceptionable breed of cocks, to detect with anxious attention even the slightest traits of change in the constitution: it may be traceable in a variety of circumstances:

1. From having bred too many years in and in.
2. From a serious loss of blood in fighting.
3. From unknown injuries in the vital and other parts, highly injurious to health.

From juvenile contests they likewise become frequently faded. And from stags to cocks they meet with such injuries as are seldom got the better of.

Many cocks are frequently bred from, without being fought. An unfavourable change with some has taken place from this circumstance: therefore to avoid any possible introduction, however trivial, omit the use of him for a brood cock, and take a blooming unmolested brother in his place. At certain seasons I have known that contagious distemper, the roop, make strange ravages. There are few, who have been breeders of this noble bird, but are well acquainted with the disorder, which issues from the nostrils and incrusts the tongue: the livid taint brings on a noxious blood—their plume fades—and their wings flag—with gaping and wheezing throat. A fever generally attends this disorder, and they are remarkably craving for water. Whenever these appearances are observed, (for its baneful mischief spreads with rapid pace) eradicate the sickly part from the sound, and give vent to the rank effluvia which often infects their eyes with frothy streams.

The under-mentioned recipe has been esteemed a sovereign balsam, nay even a specific :—

> TAKE—Green rue and sorrel, cut small, each half a handful.
> Celadine, half a handful;
> Flower of brimstone, sufficient to form the mass, with a quarter of a pound of fresh butter;
> Made into pills about the size of a small nutmeg, giving one for three nights together.

There are several compositions given, but I know of none equal to the above.

Adverting to the change in constitution. Where your fowls have enjoyed a series of blooming health, and in all other respects unexceptionable,—but nevertheless in the routine of your breeding a sudden change takes place, their well known character would induce you to investigate with the strictest scrutiny, from what part of the brood it originated; and if a regular account has been kept, this will not be difficult to ascertain. If they are derived from several sisters, with one cock, it may be difficult to detect from which of these sisters the diseased bird was produced—supposing the change to be only partial and not general. The sire of course claims your first attention. Mark steadily his exterior to see that he retains

> The standard ruddy bloom of health;
> That his feathers are not dry nor loose
> But mellow in the feel, bright and firm;
> His flesh firm and compact;
> His legs well under him, and his crow clear and sound.

These enjoyed, we would suppose the taint is not in the cock; but looks and feel are frequently deceptive. Trust not therefore immediately to appearances, for unsound cocks will shew themselves when feeding and in their fight — therefore as far as regards the cock let that be your ultimatum.

On the investigation of the hens, it is necessary to call to mind, if any of these sisters exhibited any symptoms of decline during the season these diseased birds were bred—yet probably not then very apparent—still some circumstances may be brought to recollection to enable you in part, to fix upon some one or more of these sisters, whose deportment and exterior might create suspicion, and lead to detection. But if no such derangement occurs, you have then to scrutinize each sister (supposing the whole to be living) with all possible attention, to see if they are in possession of all those requisites that constitute health; such as those detailed for the cock, with these additional remarks, that hens upon a decline have always their ordure profusely attached to their feathers behind, the plume faded and dry, the body emaciated, the head wan, and the eye watery.

If such are found, and the cocks prove to your satisfaction, we must conclude it arises from those such symptoms were found upon—and the rest of the sisters being undeniably in health—and your change only partial—we may safely conclude that they would produce a healthy race—and those remaining brothers and sisters the change was found in, are to all intents and purposes inadequate to produce every wished-for excellence; and I have found from repeated trials, by adopting the following mode, that a cross in this case was never necessary, but frequently robbed them of some of those rare qualifications they had uniformly enjoyed. A general failure wants no cross, but total eradication.

To promote the means of keeping these fowls in their wonted health and valuable acquisitions, if you are in possession of stags and pullets as well as cocks and hens, select the blooming of them both, and put stags to the two-year-old hens and cocks to the pullets, and you may for years continue in the same blood, observing to put them invariably together in the above recited manner, never on any account making use of either cock or hen after their being two years old; for by adhering rigidly to present youth to youth, you can scarcely fail of success. It seems then that nothing is wanting but to discriminate with judgment in selecting those for this purpose, and to keep a watchful eye hereafter in their progressive growth.

I know of no system in breeding so big with error, as that of the general propensity of crossing. Nor do I think it warrantable, unless your present brood of fowls is wanting in some required excellence which cannot be deduced from themselves; for although health is the grand desideratum, yet they may be deficient in heel, which in a game cock is so essentially necessary, that unless they possess it, we consider them of little value, let their other qualifications be ever so great. To accomplish this, a cross becomes absolutely necessary; and such a one as may not only contribute to that end, but also be similar in feather, constitution, colour of the legs and beak, as your own; and by such a choice you bid fair to have them regular. The procuring a hen or cock every way calculated to meet your wishes, may be attended with some difficulty, but they are to be met with either by interest or money. If you obtain a cock put him to your own hens; but if a pullet or hen then to your own cock. I should prefer the latter to the former, so much so that I should rather give two guineas for a pullet or hen, than I would one for a cock.

If you are fortunate in the produce, decline breeding a second time with your own cock, but with a stag of that produce next year with your mother, and on the following year the same stag (then a cock) to pullets of this second produce. You will then have three-fourths of your improved cross, and you may follow the same method as laid down for the continuance of breeding under the change in constitution.

When we come to consider how long cocking has been a favourite diversion in this island, it becomes a matter of surprise that refinement in breeding has not kept equal pace with many other departments in the sporting line—more especially when it is known that gentlemen of the first abilities have ever been engaged upon the sod. But there are exceptions to this remark: for the present day speaks loudly of the fame and renown of the Cheshire cocks, and I should hold myself wanting in due deference, if I were to pass over my acknowledgments of Lord Derby's cocks, whose achievements and regularity exhibit sufficient proofs of his lordship's well-founded system and attention. But my allusion is upon general breeding, which has but little claim to the character of scientific amateurs.

It is a task of no common delicacy to please so numerous a

class of breeders, who are so wonderfully bigotted in their own practice, still I hope they will not be offended with the frankness of this remark—and if the general part of my readers will benefit by the perusal of my own, it will be a flattering result.

Remarks.

I know not if the art of breeding exceeds that of feeding; though I am well aware that the latter, like the former, has never yet attained those scientific principles so as to enable feeders to fight equally regular and well; for how frequenty we see individuals in this profession out-strip their antagonists beyond the bounds of that proportion a scientific man should labour under, though equally well cocked, and enjoying every mutual and local accommodation. In the next meeting the same feeders, with the same sort of cocks, and the same antagonists, shall experience an equal, if not superior defeat—whence arises the uncertainty of cocking.

I am aware that in all professions you meet with some more eminently gifted than others—either by the advantage of superior education, or by a long course of practice. But gross and serious errors must, in the end, prove their inability to become favorites of the public. I do not mean by this to reflect upon any individual feeder, or the profession at large; for I well know they have to contend with difficulties, not easily surmounted; for a feeder who comes a thorough stranger, where he has not been in the habit of feeding, and equally so to the constitution of the cocks he has engaged to feed—his knowledge of course must rest upon appearances merely of a superficial nature, not equal to the test of experience, which enables him to provide for them with confidence, and he knows how much they are capable of turning off without waiting events;—in fact, the more experience he has with his employer's cocks, the greater probability of success: for they have to combat with various constitutions in feeding, and frequently dissimilar consequences result therefrom. A prior knowledge to what extent they can bear reducing, and to what degree of facility they are raised, are beneficial acquirements, and must preponderate over an entire stranger. It is not much to be wondered at, that a feeder under such a disadvantage should fail in his first attempt with an antagonist, who has been in the habit of feeding the same cocks for years. These failures frequently take place, and are as frequently the consequence of changing feeders. In the six great mains which I fought with——Cussans, esq. at Loughborough and Derby, John Beastal fed for me, whereas Mr. Cussans had a fresh feeder every main: the result was that Beastal won five mains out of six—from 11 to 7 a head. Here not a feeder but David Smith had any previous knowledge of their cocks—cocks too of his own breeding, and were the well-known Greys; but my Black Reds under the many year's experience of their feeder, seldom or ever failed of success. A propensity to change under these circumstances (more particularly if your feeder ranks as an able professor) should not be encouraged, because he has failed in his first attempt: but let him feed your trial cocks with as much attention as for the main; give him all the local knowledge of them that you can, and let him be acquainted with every character of your fowls, and you will find him superior to any change you generally can adopt. You will not find yourself much at a loss, on his progressive fighting, to determine your future choice; he will either exhibit some trait in favour of your initiation, or he may drop into mediocrity.

I have remarked from several years' fighting, that those feeders who have advantaged themselves considerably a-head on the first day's fight, were frequently a-back on the second; when those whose cocks were advancing to perfection for the second day, outstripped their antagonists by many a battle a-head on that day, whereas had they produced their cocks nearly equal the first day, the mains would have been much better contested than they were: that is, the majority in the mains would not be so great. Why each day's fight should not be co-equal, I am yet to learn, for there is a proper time allotted to their correspondent weights;—they should be up to their condition on the respective days.

It is not intended that a pen of cocks should have three gradations—indifferent the first, better the second, and superior the third,—but the cocks are set conformable to their weights, and on those very days they are calculated to be up to their fight.

Let us admit that one of the two pens of cocks may be superior, and the feeders equal in ability—that superiority will prevail more or less—but if there is an error in judgment they must suffer. I do not mean to enforce that feeding gives every exclusive advantage; for it is well known that cocks shall be equal in game, constitution, and diet, yet if bad heelers, the odds must be greatly against them, for it will be predominant; but when we see birds of equal stamp so unequal in their fight, my readers would not be at a loss to say what was wanting.

REMARKS

ON

Steady Breeding, etc.

The well-known Cheshire Piles have always been favorites with the gentlemen of the sod; and for reasons most urgent, for you have in them every thing that constitutes good fighters: they are lofty, well shaped, good game, uncommon heelers, and perhaps their feathers as little objectionable, taking them in the whole, as any Piles in the kingdom. The breeder of these birds has continually, and I may say invariably, for many years, pursued a steady and uniform conduct in keeping them to themselves. Even the light Piles have been kept separate from the dark, and by judiciously putting them together he has established the well-known reputation of these cocks.

Mr. Rilands has not split upon the rock which many breeders have, and still continue to do; but steadily endeavours to amend any deficiency in these fowls, by a well managed choice of the whole. Many attempts have been made in different parts of the kingdom, to cross the Cheshire Piles with similar feather, but I never yet learnt that they were equal to the original Cheshire. I have taken much pains to establish, what I have ever esteemed valuable cocks—but they never produced any thing equal to themselves, notwithstanding the hens that were put to them were the Beverly Piles, and by no means inferior to the Cheshire. Perhaps their local situation may be more conducive to their constitutions than any other country, and I am fully persuaded that that circumstance contributes very much to their excellency.

We have but few instances where a set of cocks have been so regularly kept and continued as these, and probably few have fought with more general eclat and success. To the prevailing instability of too many breeders may justly be attributed the very few, who can boast of being in possession of so regular a set of cocks, as the above recited. There are a great many juvenile breeders in the county I now reside in, whose engagements in cocking are considerable and frequent, and, if I may hazard from report, are exhibiting cocks high in repute: they are too, in that situation in life which can command all the necessary requisites to establish whatever may be deemed perfection in cocks, and if steadily pursued, I doubt not but will be accomplished.

This idea does not arise from a solitary instance with myself for I have experienced a degeneration to a greater extent than that of the Cheshire Piles. In the years 1785 and 1786 I bought of the Rev. Mr. Brooks, of Shiffnal, Shropshire, thirty pair of cocks, that were brought from thence to Loughborough, and I believe no cocks ever gained a higher reputation than they, through the whole contest of three days, with advantage of nine a-head; the cocks pitted against them were selected from the first amateurs surrounding Loughborough. Our cocks had sustained an injury which we believed at the time to be serious; but Beastal's never failing cordial (which might be called a grand specific) renovated all their finer faculties, and spurred them on to victory. Many were the applications for the purpose of breeding, and on that year as well as the following I never bred more, having put them to hens my Reverend friend had sent me with the cocks, as well as to my own that were similar in feather, &c. The whole of these bred in 1785 and 1786, kept up their character, and we had every flattering hope of enjoying all we could wish for in these cocks; but to our great surprise and disappointment, those bred in 1787 and fought in 1789, exhibited such a falling off, that very much deranged our future breeding—nor could we ever recover their pristine excellence—as there was no prognostic of any fade or change in constitution, to rouse us to any trial, or any thing to disturb that high confidence reposed; so of course they were fought without any trial, but I am well aware that we ought not to be governed by the most flattering appearances, or past merit, but give them a fair trial. How far that can be accomplished with strict propriety calls for the aid of practice to point out the particular seasons, most natural and congenial to make trial of cocks that are to be fought in March and April—which were the allotted months we fought in—and if it were improper twenty-six years back, how much more so would it be at this juncture, when our springs are much different, partaking more of winter than spring. Of course our cocks are considerably backwarder, that any trial betwixt casting and perfecting must certainly be decided against [a trial when stags, unless you fight stags is not to be depended upon]—but take them between cocks and stags when in full feather, and I doubt not but at such a period you may abide by the result. From the very hour that a change in the system takes place in a cock, previous to molting he has more or less fever; and in that progressive work of nature fever has always a share in the act; this must operate unfriendly upon the constitution, oppress and flag his natural gaiety, and render him unfit for vigorous action, and so in a greater degree in the next state of renovating his feathers, and therefore unfit for the purposes of trial.

It may be thought superfluous to give a detail of our proceeding with these fowls—except briefly to say that every practicable attention in all the various departments in breeding were bestowed to hold them to their original excellence—but proved unavailing. Our endeavours then to develope this extraordinary deviation were unremitting, and every enquiry and investigation were made use of to elucidate and aid our opinions, what might be the cause which had wrought so serious a change in such valuable fowls—the aggregate of which only afforded a solitary surmise that a removal to distant country was the cause. Before I quit this subject I cannot refrain observing, that several of these cocks up to three years old were fought in different mains, three or four times, with success—and one Ginger-Red in particular fought in five mains in one year, twice at Burton-upon-Trent, Staffordshire, at Lichfield, Derby, and Nottingham, without apparent injury: this cock's exploits will be found in full in the first number of the Sporting Magazine. Of all the variety of cocks in upwards of fifty years extensive knowledge of them, I never yet met with any that were equal to the never-to-be-forgotten Shropshire Reds.

ON

THE NECESSITY OF MAKING

Trials.

The mode of trial varies as much as any circumstance in the routine of cocking. Those who breed but few, implicitly rely upon them from one period to another: others content themselves with a trial of stags, or from some imperfect brother, taken from his walk and immediately devoted to proof—perhaps pitted against a cock considerably heavier than himself, contesting a struggle under every disadvantage,—and this only to know if he is good game. These trials are too incompetent, and too trivial for any number, much more so for a set of cocks, where a main of any consequence is depending. It is now found highly expedient with those who have various sorts, not to hazard a match without a regular private trial, under the management of those who feed it—and from such fair cocks from whose actions they may form a proper judgment of the whole. In this trial they ought to be fed with the same caution as those for a main, abating no one circumstance whatever; and where a great many are bred, they may be so selected, as to match them sufficiently accurate, which will put them in the same situation as those who are meant to be depended upon; and if any suspicion arises from such trial, if those are brothers, let the survivors be cut up, giving them every equalizing chance through the whole contest. If this is not complied with, it cannot be esteemed a fair trial.

It may be necessary to explain, to some of my readers, what is to be expected from a fair trial. Cocks proving good game denote only a part of their excellency, for they may be deep game, and yet be deficient in what constitutes the best and first quality of a cock—that of being a good heeler; and if thus wanting they must be deemed imperfect. They should be

>Close hitters,
>Bloody heelers,
>Steady fighters,
>Good mouths, and come to every point.

A hasty fighter indicates a desire to get shut of his cock—and if his antagonist gives him too much trouble, he will frequently give him ground for agreement. It is also a bad sign to see him crouch and give his head away: nor is a shifty cock to be admired, notwithstanding he sometimes win.—There are cocks, even in the extremity of death, that will shew you the goodness of their intentions, so much so as to render themselves dangerous.

To judge with nice discrimination the various effects the piercing heel has upon the frame, more immediately to catch the momentary gash upon the vital parts, to watch with cautious eye how he bears the ebb and flow of departing existence—as in this gasping stage (if I may refine upon the sentence) their heroic fortitude is frequently conspicuous, and their last moments are a trait of the greatest magnanimity. On the other hand some cocks would display an early propensity to shrink from the sod of danger; for as there are degrees of cowardice, so are those degrees variously exhibited—and many a coward, if they had the power, in their fleeting passage would quit the field of battle. He, whose interest it is to mark and to commit to paper this scene of combat, gives due merit and applause to such meritorious warriors, who have stood foremost in all their various exploits, that the surviving brothers and sisters may hereafter do justice to these departed heroes, and hold up their fame and glory to their allotted extent. When I am speaking of various degrees of cowardice, so are they variously urged by pride and jealousy, the least alarm of either hurries them on into that impetuosity (more remarkable in some than in others) which with them is not to be appeased but by death.

A trial holds to view, probably all that is valuable as well as defective; for they exhibit as many modes of defence and attack as the pugilistic tribe—and their blows have more or less of the destructive, according to their better art;—a wide striking cock seldom carries death with his heels; if his legs are out of the direction of his body, so of course they are further extended, and they are generally esteemed dry spurred cocks—and it generally happens that they are a low flying fluttering cock,—but on the contrary, if his legs are in a due direction to his body, he is more erect, rises higher, and is a close hitter,—his battles are generally short.

Brothers will not always prove equally good; yet if close bred, from a regular set of sisters, any little apparent want of constitution should not suffer them to run away.

A few years ago, some friends thought it necessary to make a cross with a set of favourite hens, then well known by the appellation of the Old cross-marked Sort. They were given to understand that cocks, similar in feather, &c. and long in favorite repute, would fight at Nottingham. These consisted of six full brothers, fine, lofty, boney cocks; four out of the six fought, and in such a style of excellence, that, exclusive of their previous recommendation, it was thought prudent to purchase the two unmatched.

Merely saying that they fought in a style of excellence is not doing that justice to my worthy friend, who pointed out their superiority, and whose judgment and recommendation gave a decided preference in their favour: I need only to say that they followed closely to the heels of our favourite Shropshires, and to them we refer our readers; and like them they never produced their equals. Every thing which could contribute to the welfare of breeding, was bestowed upon the brood, and their progeny gave us such flattering hopes, that we were undetermined if we should give them a trial when stags, but the produce being very numerous, and in the midst of our annual great main, it would have been placing that confidence, experience urges me to say we ought not to have—therefore on mature deliberation a selected number were committed to the management of our old feeder Beastal—fed at the house of one of the gentlemen concerned, and enjoyed all the advantages of cocks devoted for a regular main, and from this cautious selection were matched with great equality—every thing was conducted with that degree of regularity which would have done credit to the first mains;—the setters had equal merit—due attention was paid to them in their intervals of fighting—and all the laws of cocking were rigidly adhered to. I have seen many mains, and a great many trials of stags, but I never yet saw any that exhibited such proofs of every excellence, as these stags manifested throughout every stage of trial. If a trial of stags could be a certain criterion of their goodness in a future state, as cocks, how pre-eminently high we should have ranked, in our endeavours to meet our numerous antagonists, with such raræ aves,—the consummate superiority must have carried every thing before them. Highly gratified as we were in their juvenile state, we were also intent upon another proof ere the main came on: the same attention was paid to them in cocks, as to the stags, but what a falling off was there; such as would induce a man to decline any attempt at future breeding,—their constitutions were so glaringly changed, that all our hopes were baffled—our arrangements disconcerted—and we were thrown upon the public for a supply which in the end was the loss of the only main out of six fought. Such a carnage ensued as seldom takes place, for the whole were made away with—and the surviving old cocks, in their trial, but too plainly pointed out to us the failure we experienced in their sons. My own died rotten on his walk, the others, though struggling under a bad constitution, evinced evident proofs of their innate excellence. A brother to the brood cocks, after being struck blind in the first fly, and won his battle, was bought at a high price by the late Mr. Brooks, in London, whose knowledge in these birds ranked high with the gentlemen of the sod;—he put him to some of his Spitfire hens, that enjoyed all the flow of health, still the progeny was equally unfortunate. These cocks had been known and fed many years by Beastal—and his annual remarks upon the superiority of them were not lost upon his employers. Such failures as these are unpleasant drawbacks in pursuit of cocking, and have their serious attendants. Our references relative to this unfortunate change was easy and satisfactory—that the breeder of these cocks had kept them in and in for so many years, without that requisite caution of putting youth to youth, was the rock he split upon—and hurled destruction on his favourite brood.

These details have for their object—

To point out the liability of change in the constitution, from a removal of cocks from a country they were bred in, to a distant one:

The uncertainty of crossing to advantage:

To shew that a trial of stags is not to be depended upon, and the necessity of trial when cocks:

That cocks may hold their constitution until they are two years old:

A failure in the constitution at three:

The impropriety of keeping in and in without putting youth to youth.

INVESTIGATION

OF CIRCUMSTANCES MOST LIKELY TO PRODUCE A

Bad Constitution in Cocks.

The well-known axiom, "that prevention of disease is better than cure," cannot be more properly enforced, than after the following recital of the various transitions in the cock from health to disease.

I do not feel myself liable to contradiction when I assert that nothing is more common than to breed from a cock, after having fought not only one battle, but several; the more battles he has been engaged in, the greater recommendation to make use of him as a brood cock. It would seem, upon reflection, unnecessary to dwell upon the impropriety of such a choice—but the mischief is of such magnitude, that it would have been unpardonable to be silent upon the subject. When we come to consider the position for combat, is it supposed possible that they should avoid giving and receiving a stab, more or less injurious; for they have no shield to ward off the deadly weapon from the vital parts; the interstices are numerous; bone, muscle, tendon, are all liable; vessels may be divided, or perforated; parts may be wounded, lacerated, or contused—and probably seldom restored to healthy actions:—the quantity of blood lost or effused, may of itself ruin his constitution. Under all these liabilities, who would build his hopes upon such a crazy fabric; admitting without a doubt that such cocks are made use of, are we to wonder at meeting with so many that have delicate constitutions.

The next, and not least of the causes, is that originating from the effects of molting. We know from experience that cocks, as well as hens, in particular years suffer such difficulties in that wonderful act of nature, inasmuch as to render their lives precarious for a length of time. It is well known that a fever, more or less, is ever attendant on this act of nature, and I have frequently observed that this operation is sometimes arrested, and remains for a shorter or longer period in a dormant state, without exciting any perceptible commotion in the economy—and when in this state death generally ensues—or it leaves them in a debilitated and faded state. At other times we see it rapid in its progress, and the ordinary course of nature enjoyed and perfected.

If hens and cocks have not completed this renovation, and fully arrived to their health and bloom, both the one and the other are unfit for the purposes of breeding.

I have remarked for several years, that if either my hens or pullets should be deprived of sitting when nature prompts, they do not molt kindly in the approaching season; how far and in what manner this resistance to nature operates on the constitution, I am unable to divine.

Cocks fed in the month of March, cut out of feather, and not matched, tendered by the mode of feeding and close covering, turned out in the middle or latter end of the same month, and perhaps the season as inclement as any part of the year,—what must the sufferings be of cocks so exposed?— it so far operates upon them that instead of being lively and blooming, they are dull, heavy, and listless,—for as yet, notwithstanding they have been up and fed, nature had not then completed her work, for not one cock in ten is fine at that juncture—and the instance of turning out, would, without a doubt, procrastinate the act of progressive nature, and leave him unfit for the purposes of breeding, or in fact for any other where much was depending. The practice of fighting mains in March, as they have presented themselves for many years, seems to me to be unnatural and premature; but this must be submitted to others, to judge for themselves—for in this chapter I am only intending to mark out the causes which have retarded, or accelerated the progressive work of breeding.

They say " of evils, we are to choose the least ;" and I think it is a much smaller evil to feed and to fight, when cocks are in full feather, and in high health, than to take them in the state described. It is not only unnatural but you are not giving this noble bird an equal chance with the other animals we select for our sport.

To the catalogue I have already commenced, I must not omit the following addition:—That of the severe injury stags meet with (particularly the forward ones) in the months of November and December—so much so as to arrest their growth, and frequently to leave a taint upon the constitution, that renders them unfit for breeding or fighting:—when stags of this description go out, it would be right to make a memorandum of the injury they sustained. A variety of circumstances are here adduced, in order to point out to my readers the necessity of being very guarded and circumspect in every department appertaining to cocking, and particularly those whose engagements are numerous, and who fight for large sums. We know that notwithstanding all our exertions and endeavours, they are not to be always crowned with success—but the discerning sportsman will not be at a loss how to account for the failure, and will give due credit to meritorious cocks.

Remarks

ON SOME FAVORITE BREED
OF
COCKS.

When Sir Charles Sedley and Hugo Maynell, Esq. fought their long mains, I believe no contests in the annals of cocking were marked with such general eclat. Their characters as judicious sportsmen stood conspicuously and deservingly high. They fought for large sums, and backed their cocks with such unrivalled spirit, that before or since, no betting has been in equal competition. Great confidence was placed in their cocks, and their feeders were upon a par,—under such circumstances, high betting is ever the consequence. The mains were upon the decline previous to my becoming a breeder, but a considerable number of cocks and hens were left in the neighbourhood around my residence. As they were relinquished, and become the property of the possessors, I was in the habit of purchasing a number of both—particularly from one of the breeders of these fowls—those that were fought were equal to any that had gone before them, and as a young breeder, I thought myself singularly fortunate in having to commence my pursuit in cocking with such well-known valuable fowls. My selected few, as brood fowls, were put down to one of the sweetest retired spots my neighbourhood could boast of, and comprised every accommodation that could render them secure, quiet, and healthful — and no close breeding could be superior. Under these local advantages, early in the spring (and the springs then were congenial and delightful) I had three hatches produced so regular, and altogether so promising, that I could not help anticipating a future pleasure in the progressive growth of what I considered valuable birds—and no young cocker was so highly gratified in having at hand seventeen stags and twelve pullets, not a feather in the whole that did not challenge their originals. So pleasing was the recollection of these birds even at the distance of fifty years or more, that I omitted prefacing this recital with a description. They were a clear Mealy Grey, nearly white breasted, without spot or streak, about four pounds six, to about four pounds eight ounces — high standing, boney, and black legs — close feathered, short hecked — small snake head, and full dark eye — their walk easy, firm, and majestic, and their breast gracefully prominent, their shoulders broad and up, and their body gradually tapering to the tail; there actions were in unison with their shapes. To enumerate the superior qualities of these cocks, would be reciting again those of the Shropshire, and like them, when crossed, they lost their original goodness; they became soft feathered, and partook of all the variety of the Greys. Had these birds enjoyed a judicious cross selected from those that were left—I have not a doubt but they would have been excellent for many years. The only cocks that bore any similarity to those, were bred by Mr. Hopkinson, and what few I saw of them fought at Nottingham. I, at that juncture, entertained an idea that they were a part of the same family.

Unfortunately, three months previous to my cocks being two-year-old, I was abroad, and was nearly two years and a half absent; of course they were entirely lost to me, for on my return they were differently disposed of.

The Black Cocks of Lord Here

Were introduced by Mr. Thomas Wilson, of Burton, Staffordshire, my place of residence, and I know of no amateur so eminently qualified as he was to do justice to any breed of cocks, however high in repute — few professors in those days were so systematic in their mode of breeding, nor were they likely to lose their former reputation in the transfer.

They were a perfect jet black, gipsey faced, black legs, rather elegant than muscular—lofty in their manner of fighting—close in their feather and well shaped; such was their established character, that wherever it was their lot to fight the odds always preponderated in their favor—in fact, level betting never followed their contest; and for three years these cocks maintained a decided priority over most in the circuit they fought in. An opposite interest in some mains operated as a bar to our usual intimacy, which deprived me of being personally informed of the alteration of these cocks, that were very conspicuous the two years ensuing,—a diminution in their appearance, their feathers long and dangling, their fire much abated, and so much unlike themselves, that they were more like crows than cocks. For reasons above stated I could not learn the cause from himself of these material changes in such valuable cocks, but if hearsay authority could be depended upon, Mr. W. had persevered in breeding from old stock—the result a certain consequence of derangement. That this error which had so frequently happened to others with all its train of mischief, should have escaped the active penetration and vigilance of Mr. W. who was sensibly alive to the refinements of breeding, has ever been a matter of surprise, for whenever it has crossed my ideas I have been at a loss, conclusively to account for it, otherwise than by a rooted infatuation, not easily to be conquered, of that too obsolete idea—that old favorite cocks may be continued for the use of propagation for a series of years, with the same propriety as the first year. If analogy will bear me out, with reason's aid—that a man of sixty and a woman of forty-five shall produce as fine children as a man of twenty-five and a woman of twenty-two, then the impropriety falls at once, and that we have to refer to the catalogue of other causes enumerated in this work. These conclusions are promulgated from well-known facts, arising from extensive practice, and for the better understanding of which, I have given the following solution:

Comparison.

Stag 1-year-old with hens 2-years-old—man 18, woman 22.
Cock 2-years-old with pullets   man 25, woman 22.
Cock 2-years-old with hen 2-years-old—man 25, woman 22.
Cock 8-years-old with hens 8-years—man 40 to 50, woman 45.
Cock 4-years-old with hens 4 years—man 50 to 60, woman 45.

From five to six, the cock should give way to his sons, and the hens to sit and nurse; admitting this axiom, allow me to express a hope that the time is not very remote, when the practice will be done away with,—that the contrary will be universally adopted by discriminating breeders,—and that unfounded prejudices and vague infatuation will vanish before the accumulated evidence of the advantages arising from a judicious practice.

These cursory details of the different breed of cocks I have adduced, in order to impress upon my readers the importance and necessity of being strictly cautious in breeding from the same fowls, without a due regard to youth: as well as when a cross is necessary with those that have no affinity. To characterise the appropriate feather and other concomitants that assimilate nearest the originals you intend to cross with; to blend the constituent qualifications to produce that uniformity so much admired, and which gives that perfection and superiority over that system an incautious breeder pursues ever to his disadvantage, has been my principal aim in writing the foregoing pages.

Cockfeeders.

Whoever of the profession may peruse these observations, will take the motive for the deed. I have suffered materially in a match of as fine stags as ever were exhibited, by the improvident neglect of my feeder, in not airing his room and pens. Even those that are frequently used for that purpose should not be exempt from this caution. Therefore it is a duty every feeder owes to his employer, as well as to himself, to see that the room is thoroughly cleaned and well aired, and that the pens are removed from the walls, taken out of the room, and well scraped and washed, and that they are as dry as the air can make them, by being exposed to it for at least eight or ten days previous to their being brought in,—that they may be properly and duly exposed to the warmth during the time the fire is continued. Pens contract a fustiness from being a short time out of use, more particularly when vacant the greater part of the year—so as to render them disgusting, and brings on nauseas and distemper.

To see the lesser minutiæ properly prepared is highly necessary; such as the straw, that it is clean thrashed, sweet, and very dry—and I should strongly recommend the use of distilled soft water, or at any rate the best soft water filtered that can be procured; and your barley thin skinned, old, white and dry, and deprived of the ends, as much as thrashing in a bag will effect: by so doing it will digest the easier. As to the other articles of diet, they must be left to the management and discretion of the feeder.

Coverings, according to the season, are used thick or thin, whichever of the two let them be remarkably clean, sweet, and well aired: I should prefer a regular set for the purpose, to be the master's property—I have seen very improper coverings. If your room is sufficiently large, and you have a regular set of pens, never suffer them to be fixed up to the walls, but put in frames with upright standards in the centre of the room—that is, double pens not directly opposite to each other, but one of the opposites a pen distant from the other. Cocks should be considered nearly stationary in their pens, they cannot avoid surrounding effluvia, and as walks imbibe much moisture they are long ere they emit it—and of course the humidity is imbibed by the pens. If the pens by being stationed in the centre become liable to a partial air, I have known double doors made use of as a seclusion, and equally so for security—the fewer visitors the better, and as few handlers as possible. Cocks may be said to be immured in a prison, when in their pens—deprived of their natural air and exercise, debarred the enjoyment of proudly escorting their hens, in culling and selecting whatever contributes to their health—roused and ruffled by noisy and repeated pride, until he becomes jaded and hoarse, and shy even to his scanty meal: use reconciles him to his prison, and he becomes familiar to the offers of those artificial luxuries which warm and exhilerate the system, and help to soften the tediousness of confinement and intrusive handling. I think it is almost unnecessary to say any thing upon the subject of cleanliness in every progressive step; many little attentions are so many gradual promoters to their welfare. An industrious turn straw has it much in his power to contribute thereto, and in the department of shaking and raising the straw the dung falls to the bottom—and of course, must by heat become offensive; surely it is no very difficult undertaking to thoroughly cleanse the pens, not only of straw, but of every apparent filth—and the better to accomplish this, have two spare pens, marked Nos. 1 and 2, equally and delicately clean with the other, one at the top, the other at the bottom—and these two pens will answer the purpose of your shifting the whole from top to bottom, until they are all finished; and this may be done with the least possible trouble: two people may go through this necessary and cleanly operation without any injury to the cocks—the enjoyment of fresh straw every day must add much to their comfort. Several pens may be emptied in a large twig basket, in preference to emptying them upon the floor, which would create dust, &c. and so on to the end. A cock-feeder in London very high in esteem, every morning made use of a pint of the strongest vinegar he could procure, and with a large sponge washed the two divisions of each pen which the cocks fed through, and which he thought prevented any disagreeable consequences arising from the multiplied exhalations of the whole. This sponge appeared to me to be a very useful applicant to cleanse and absorb unnecessary moisture. Another excellent accommodation was that of hospital pens adjoining the room he fed in, these were also very clean and well strawed, for the purpose of those cocks that won, and the same number put upon the hospital pen as the cock fought out of: so that they were easily recognised, and the feeding pens by this method were kept pure and sweet. Pens might be constructed so as to take to pieces, cleaned and kept dry until wanted, and put together at a small expence, which I should recommend as preferable to all others.

Cocks are frequently brought up from their walks in damp and filthy bags, which is a bad practice, and never ought to be suffered. We cannot be too cautious in guarding against every possible circumstance that might introduce any unfavourable disease in the hens; such has been the case frequently, and instances have been known from villainous motives.

In respect to feeding, it is a province I never encountered—nevertheless I have paid much attention to the different modes each feeder made use of, and although varied in several respects, yet each have fought with good and bad success in their respective modes. It requires an extensive practice to form a complete judgment of the condition of a cock on the day of fight—much depends upon a good finger, and a knowledge of that position a cock puts himself in, in and out of condition, exhibiting different feeds and forms; it is not the feel that feeders are in the habit of using to know if a cock returns his meat, but that feel which gives to the grasp that firmness, compactness, warmth, and fire, all of which are felt and seen:—the eye pourtrays the rest of his condition.

The firmness of his flesh indicates health and good feeding. The contraction of his legs to his body, vigor and heat; and a cock under these circumstances, shews what is termed full of fight—the contrary is soon felt; for he is

> Soft and heavy in hand,
> Legs low and dangling,
> Eyes dull and unmeaning,
> And his whole cold to the feel.

When a cock arrives at the top or height of his feed, art can go no further, and when it takes place the day he fights much credit is due to the feeder; for cocks retain that zest but a small time, and become retrograde every day after. All cocks when set, should be so ordered, that they should arrive at their proper fighting condition on that very day they are to fight—if either under or over—and his antagonist fortunately otherwise—it is at these critical failures that one side is so frequently cast in the back ground.

Cocks vary much in their mode of fighting;

> Some are hasty and fiery,
> Others cautious, wary, and close hitters,
> Some wide and generally dry heeled,
> Whilst many are lofty and darting.

Those that are low and fluttering, are seldom dangerous in their heels; the latter description are those that are destitute of that tapering shape which so eminently distinguishes them in their superior mode of fighting. Cocks that are as broad behind as before, have their legs thrown out of the line of the body, and of course are wide in their fly and dry heeled.

To judge well of a battle requires much attention—a quick discerning eye—a knowledge of those parts of the cock most liable to sudden and destructive fate, and which turns the fluctuating tide of odds against them; others that are more slow in their effects, yet fatal to their victory—many are momentarily crippled and yet not immediately detected: variously are the heels directed, and many parts are perforated with little injury in the heat of battle, although felt when cold—these are not alarming to the adept, and they take advantage of those who are:—a cut throat is for the most part very conspicuous.

A well-known amateur describing a battle has the following lines:—

> "Now hostile rage each daring foe maintains,
> "And death, as fate inclines, alternate reigns;
> "In various shapes the missive blow appears,
> "And dire destruction 'midst the conflict bears—

"Now purple life unloads the turgid veins,
"And gushing down the crowded circus, stains,
"Or stagnates, swells the throat, and vital air restrains."

Copy of an Article for a Cock-match.

Articles of Agreement made the ——————— day of ——————— 181   between W. S. and J. C.

First, The said parties have agreed, that each of them shall produce, shew, and weigh, at the ——————— Cockpit, ——————— on the ——————— day of ——————— next, beginning at the hour of seven o'clock in the said morning, ——————— cocks, none to be less than three pounds six ounces, nor more than four pounds eight ounces; and as many of each party's cocks as come within ——————— ounces of the other party's cocks, shall fight for ——————— guineas a battle—that is, ——————— guineas each cock, in as equal divisions as the battle can be divided into, as pits or days play, at the cockpit aforesaid; and that the party's cocks that win the greatest number of battle matches out of the number aforesaid, shall be entitled to the sum of ——————— guineas as odd battle money; and the sum is to be made stakes into the hands of Mr. ——————— before any cocks are pitted, in equal shares between the parties aforesaid; and the parties further agree to produce, shew, and weigh, on the said weighing day, ——————— cocks, for bye-battles, subject to the same as the main cocks before-mentioned, and those to be added to the number of main cocks unmatched; and as many of them as come within one ounce of each other, shall fight for two guineas each battle, to be as equally divided as can be, and added to each pit or day's play with the main of cocks;—and it is also agreed, that the balance of the battle money shall be paid at the end of each pit or day's play; and to fight in fair reputed silver spurs, and with fair hackles, and to be subject to all the usual rules of cock-fighting as is practised in London and Newmarket;—and the profit of the pit or day's play to be equally divided between the said parties after all charges are paid and satisfied, that usually are thereupon. Witness our hands this ——————— day of ——————— 181 .

    Witness   W. S.
    J. W.   J. C.

Rules for matching and fighting cocks in London.

To begin the same by fighting the lighter pair of cocks (which fall in match) first, proceeding upwards to the end: that every lighter pair may fight earlier than those that are heavier.

In matching (with relation to the battles) it is a rule always, in London, that, after the cocks of the main are weighed, the match bills are compared.

That every pair of dead or equal weight are separated, and fight against others, provided that it appears that the main can be enlarged by adding thereto either one battle or more thereby.

Further Remarks.

Breeding is involved in many difficulties, for cocks will partake of a variety of shapes; and, with some feeders, a cock, full in his girth, and narrow behind, is preferred; when, with others, a lofty spiring narrow cock is approved. As this is a well-known circumstance, and if you breed to any extent, you have to combat with all their various partialities, if those fall to your lot who are enamoured of the former description, the cocks of the latter become a loss upon your own hands, although the breeder is confident they have every fighting excellence that can attach to the cock: and, although I have ever recommended the establishing the first class, yet there are certainly many instances where that partiality ought to be set aside. This present year (1814) a gentleman of my acquaintance, who has been long in the habit of breeding some capital feathered Birchins, which are invariably lofty and thin; and such is the excellency of these cocks, that their average winnings have been nine out of thirteen for several years; yet he is now so circumstanced, in respect to his present connections, that not one of these birds have been an object to the present feeder. This is not brought forward as being an advocate in promoting the breed of the narrow-shaped cock, yet cocks so eminent in their blood, feather, and heel, are much preferable to those of superior shape, not so gifted. My friend is perfectly aware that his birds are deemed deficient in shape, but he still means to persevere, rather than to hazard the giving them the so much-esteemed shape required by a fresh cross, or the giving them up altogether, under the idea that he may lose in them what he esteems a superior acquisition. His unconquerable arguments are, that he had rather have them with their matchless heels, than the most esteemed symmetry without.

Feeders say they expect good shapes; but what they call good shapes do not bring with them good heels. A lofty, narrow-shaped cock is wonderfully agile in his sparring, and for the most part more dangerous in his spurs, than a contrary shape; but a broad cock, with equal share of heel, must have superior resistance and power, and, if attainable, they are, without a doubt, the most to be approved. I should think my feeder nicer than wise were he to refuse cocks of mine of that description, merely because he did not like their shapes. Sporting gentlemen, one and all, give unquestionable preference to a cock with a good spur, as the most decided acquisition appertaining to a blood cock, therefore I cannot impeach my friend's attachment to his well-tried favourite birds. His ultimatum he ever holds out to me, that he will fight a main, or any number of mains, with any gentlemen in the three kingdoms, for their sum, meeting him half-way—and such a main I am now commissioned to make. I did not mean to convey his wishes through this medium, as a season may be passed over ere this is generally out, but my more immediate reason was to imprint upon the minds of my readers, that cocks, deviating from the true standard of shape, are not whimsically to be rejected when they have so many excellencies to counterbalance. This is not a solitary circumstance, which can only be adduced in mitigation of these sometimes rejected cocks; for it is now full in my memory, that a set of cocks, the property of a Captain Barnes, a resident near Burton-upon-Trent, Staffordshire, where I formerly resided, fought a great many cocks of this description, save, that instead of their being equal in feather to the last recited, were perfect cuckoos,—in most other respects, as to shape, similar. These cocks were still more lofty in their sparring, and an adversary had seldom the chance of a long battle: they were quick dispatchers, and deep game. These, notwithstanding their feather, which, according to our ideas and modern improvement of breeding, would be sufficient to reject them in toto, were sought for with avidity, and no cocks had warmer advocates, or more general followers. These party-coloured birds are apt to degenerate in their constitutions, but that was not the case with them during their being in possession of their original owner; for I had an opportunity of knowing them for several years. It was remarked that no cocks retained that beautiful vivid red, that lustre of health so every way conspicuous in these cocks. The vivacity of the eye, and their high-beaming spirit, ranked them a superior class of birds. The attitude, carriage, or disposition of the whole body in these cocks were remarkably graceful, and their head and neck were always in proudly motion. At the disease of Captain Barnes, these cocks fell into the hands of the neighbouring colliers, became crossed with other fowls, and lost their original character. This is another instance of the great difficulty of crossing to advantage; for we see, notwithstanding every excellence, if there is not that uniformity and corresponding character in the hens, or cock, it is more than high odds if the produce is equal to either originals. If they, like many others, had been properly kept to themselves, like the Cheshire Piles, they might and would have fought their way into admiration. Adverting to the stately vivacity of these birds, it has frequently occurred to me, in a variety of instances, what a wonderful difference there is in brothers in respect to that outward and pleasing liveliness, from those that have been at hand-fed walks, and those from well-furnished walks: the former become stationary to the spot, without action, motion, or employment: he appears oppressed and heavy, and his nourishment brings on a contracted aversion to action: nature, as it were, preys upon itself. How necessary, indeed, is action and exercise to the body, may be judged by the difference we find between those cocks who labour in plenty, and those who are in the predicament alluded to. And how much superior is the complexion and constitution which labour creates, in comparison with that habit of body we see consequent to an indulgent state of indigence and rest. Several of my cocks that are hand-fed, which are at proper distances from contiguous annoyance, are never fed by those who walk them, but inviolably have their corn given them twice a day at a distance from the house, spread upon chaff; at other times upon short straw, that they may labour to obtain it. By this method they are generally upon the alert, are seldom or never near the house, and are no way inferior to the others.

This present season I have adopted a mode of breeding which should be strongly recommended to those who have proper conveniences. They are equi-distant, not more than three to four hundred yards, some two or three inclosures, intersecting each place: two of these are fixed under a high covert hedge in the form of a dog-kennel, and are well secured; the hen has a small apartment divided from the part they roost in, to lay and hatch. At each of these places I have only one hen or pullet with a cock: these are selected from sisters whose shapes, feather, &c. pre-eminently distinguish them from the rest. Each of these situations is only a small distance from running springs: they have well-sheltered edges, and a fine dry carpet. It is not the situation only which contributes to this promising mode, but a chance of deriving a greater probability of success, by having no other hen to interfere—they are the choice of your whole stud; and if any failure should follow, there will be no difficulty in ascertaining from which it originates, as a fair trial of the cock would determine: for, if satisfied with him, it must rest with the hen. When several sisters are breeding from, the greater difficulty you have in finding from which the failure proceeds—but from this mode you have a short reference. At a sufficient early period the produce was about one hundred chickens. This trial consists of two stags which are brothers, with each a two-year old hen, which are sisters—they have not a deviating character, consistent with the most approved choice for crossing; the other two cocks are two years old, and also brothers to the stags, with each a pullet, sisters to the two-year old hens, which were with the stags. By this arrangement you are breeding a great number of brothers in blood; and I never yet saw a produce so regular in feather, and every appertaining quality, as they are from this systematic mode of selection; and I have a pleasure of anticipating a fortunate result. A greater number may be produced than from such an allotted few, if you have the command of a sufficient number of blood hens—to sit in due season, which may be done by taking the eggs, as they are laid, and the nest egg when inclined to sit.

It may be said, that from four sisters and one cock may be had as many birds as devoting four cocks and stags to four hens and pullets. Admitted—but you have the advantage of four cocks to one in the mode recommended; for, notwithstanding every due attention is had to select a single cock, free from every apparent fault, yet we have frequently found ourselves disappointed either from a change in constitution, or some other cause. On the other hand, you have the best calculated chance to succeed. I have before observed, under the head of breeding, that sisters may so essentially differ in health, &c. and the difficulty of attending the prevention of laying to each other is scarcely to be accomplished; of course you cannot distinguish each from the other, and if any defect takes place, you are totally at a loss from which of the sisters it originated. It has many other claims to preference, which the cock and the hen must be benefitted by thus pairing, and the reference in case of failure much easier detected, that I flatter myself that whoever makes the trial will find the advantages so materially in favour of this mode, that he will in future adopt it with facility.

It has been a matter of astonishment to me since I have resided in Nottinghamshire, that so few good cocks are to be met with. Amongst a number of breeders to whom I have a friendly access, few are sufficiently cautious in the crosses, so essentially necessary to produce a regular set of good cocks; but there is a prevailing partiality amongst cockers which is not easily to be done away, and probably must be the work of time. We are frequently disappointed in our best endeavours, and even at a time when our hopes and expectations have run high, grounded upon the idea that our selection comprised every rare qualification to warrant success;—for I may yet venture to add, without partiality, that if they possessed every essential character that could possibly constitute a proper cross, still some heterogeneous mixture might lurk inherent, to disappoint all the practical attention and wary caution, to render the cross complete, and to establish a regular set of cocks. Is it, then, to be wondered at, the failure of those who breed without any of those nice regards, either to similarity of feather, &c. and those congenial attributes to form a complete whole?

Still I flatter myself, that the professed amateurs of this noble bird are more alive to the improvement, from the well-known superiority of cocks bred by those eminently distinguished for their abilities in every sporting department; the leading pre-eminent character they have supported must awaken the efforts of those who would profit from such enlightened precedents. Those of Lord Derby are the strongest recent proofs of my assertions: their successful prowess marks that nice discriminating care ever attendant upon whatever his Lordship pursues in the sporting line. His cocks are more regular and undeviating than most cocks within my knowledge, a certain criterion of that well-regulated system his Lordship has long and successfully pursued; and I will hazard an opinion, that, were breeders to adhere to the mode recommended, few occurrences out of the ordinary course of breeding would be less rare, and would tend to establish a race of cocks infinitely superior to the present. It is astonishing to see how wedded some amateurs are, so every way dissonant to the true principles of breeding—and my endeavours to remove such obsolete practice has ever been an Herculean task. It avails not your setting forth the impropriety of crossing contrary feathers, &c.—they are satisfied with the idea that they were both undeniable of the sort. If they are not equal to the originals, they are totally at a loss how to account for their deviation, reconciling themselves by advancing some pretext totally incompatible with the true cause—and they will even venture upon the same fowls another year. Still involved in error, they seek out for such as in their estimation will bid defiance to any possible disappointment—breed on without regard to what constitutes any congruent principle, and they become tired by their own infatuation. There are others who have enjoyed a more extensive opportunity, by mixing with their superiors, through feeders introducing their cocks approved by them; thus benefitted by repeated intercourse, have bred with considerable success. Sant, well known to the gentlemen of the sod, who resided in Derbyshire, was in the habit of breeding as good cocks as most men in the kingdom; he adhered closely to every requisite for judicious crossing, and keeping them properly together, and for many years no man fought with greater success; and I know of no cocks that were more generally sought after — a convincing proof of his attention. He had the advantage, too, of residing amongst a numerous set of men who have long been in the habit of breeding; and in the interval of his rapid success no country was in such general esteem, and any number could be procured. He was a ruling satellite over those hardy set of colliers who are invariably cockers wherever they reside; they are unwearied in their endeavours to procure such as are first in estimation; and whatever he judged superior, they were always at his command—and Sant reciprocally assisted them from his more numerous produce. He was deservingly high in Beastall's favour; who, during Sant's celebrity, was generally esteemed in his profession. Beastall had great privileges from his employers, and whatever cocks he thought would promote or add improvement to those of Sant, he was never sparing, and they were duly appreciated by him. Under these singular advantages we need not wonder that Sant should, for so many years, enjoy the well-earned encomiums of the amateurs of the sod. Beastall bought and fed many of his cocks, and of course several fought in the mains in which I was engaged. They were a very dark black-red, striped, uncommonly black upon the heck, black beak, black legs, very lofty, and fought high weights; they were favorites with Beastall, and in his hands became favorites with the public.

The high estimation these cocks were held in, caused such repeated applications to breed from, that Sant found no difficulty in obtaining sometimes very exhorbitant considerations for those he chose to part with: but the difficulty of getting any hens from him, from which you could derive the most essential advantages, were next to a prohibition amongst the parties concerned in these birds. However, I bred from several of the cocks, and the first of my trials was with some very favourite hens from Leicester, bred by a Mr. Needham there, but they proved only second rates: the hens were as well descended as the cocks, but the produce varied as much from the originals as possible. They were the choice of Beastall, who was well and long acquainted with both sorts—and such was their strong affinity, and selected by a man so every way qualified to judge of their proper essential characters, to mingle those similarities, so as to form and stamp their like—that I had high expectations from every cross I made. But such, I repeat, is the difficulty of crossing to advantage, that an amateur should be possessed of some persevering degree of patience to sustain the frequent disappointments which extensive practice will ever make him heir to. All feeders are not breeders: and one ability is perhaps coequal with the other, where all do not excel.

Cocking, like all other sporting pursuits, has its ups and downs, with all their attendant disappointments—and whatever is most predominant in our pursuits, is more or less followed with avidity and with various success. I am aware, too, that no scope of practice, however great, will make some professors go beyond the line of mediocrity. The advantages sometimes derived from those callings must arise with those who were initiated with men of known abilities, and if fortunately aided by other requisites, they are frequently known to tread close upon the heel of their employers. The public voice is loud in praise of Harry, who was long Thompson's assistant, and who attaches credit to his master's well-known abilities—his industry, and general good conduct, with his recent successes, cannot fail to recommend him to the notice of the amateur, and to bring him forward with some degree of eclat. Thompson's abilities as a feeder and a setter ranked him one of the first of his country; and as far as my judgment could decide, few men exhibited a cock upon the pit in higher style. Whether he fought a cock of high or low weights, they were conspicuously alive to that degree of perfection Thompson was capable of giving. His mode of reducing great weights was singularly effective; equally fortunate was he in lifting them up to their proper standard with every vigour and fire. It is much to be regretted, that men of rare abilities in their profession should have some reigning foible to throw a cloud over their otherwise meritorious actions. If Harry is in possession of the mode above alluded to, all the other routine of practice could not escape him. I have been well informed that Thompson enjoyed this secret for several years, without its being known to any other feeder; and as Harry was a valuable assistant, let us hope that he is not quite without it.

Matching

In general is made from sudden impulses, and frequently entered upon in that degree of hurry which seldom ends with the success a deliberate provision would probably insure—such as your being well cocked, and stags in succession, if for a term of years, a competent feeder—the distance you have to meet your adversary; and as feeders are generally mentioned previous to the articles being made, weigh the public opinion between the merits of the two, and what probable odds there is against you as to your adversary's cocks. If you are to fight upon his ground, and your distance is considerable, you will not meet upon equal terms. The aggregate of these duly considered will guard you against contested odds. Suppose your cocks and your feeders upon equal reputed merit, your distance coequal—I see no impediment to closing. Cocks that are to be conveyed to any considerable distance by a carriage, be its construction what it will, and every security made use of for their ease, &c. they still will suffer more or less fatigue; and although the action and re-action of the carriage may in our ideas be trifling, yet the effect is irksome, and the sensation in that kind of pendant motion unpleasant and jading, bordering upon sickness, and a cock does not easily regain his wonted liveliness; but when you meet on equal terms, the one is under the same liability as the other.

My sole aim in this is to guard my readers against entering upon a match too precipitately, that you may avoid meeting your adversary with the odds against you; for, however partial we may be to our own cocks, theirs may be equal, if not superior, and where any advantages take place in the match, you may anticipate a suffering in proportion to any neglect in the arrangement. In respect to time, they are calculated from various motives most conducive to a good meeting: and take the cocks as they are at the time, these probably include three-fourths of the mains that are fought—they are established to answer the sundry purposes of races at different periods of the year, and help to fill up the vacancies of the respective days. But mains fought for a certain number of years, without regard to any meeting (except the second Newmarket Spring Meeting) which is much earlier than any main ought to be fought (for cocks are at least two months later in completing their feather than they were thirty years ago) and as such, the first week in June is as early as any independent main ought to be fought. We cannot expect that cocks will fight equally well at all seasons and in all circumstances—practice convinces us to the contrary. When nature has perfected her works, and cocks have reached the summit of renovation, a cautious matcher is then giving his feeder every advantage he can wish for to exert his abilities for his interest, and he ought not to contend with difficulties which by a judicious foresight, might have been prevented.

The Nottingham Journal is just now at hand, exhibiting three days' fight at Manchester, and the result of this meeting has a strong tendency to confirm what I have advanced. The three days' fight is as follows:

HEAP.	m.	B.	HARRISON	m.	B.
Tuesday	8	2	Tuesday	8	3
Wednesday	5	1	Wednesday	1	1
Thursday	5	1	Thursday	1	1
	18	4		5	5

It appears by this statement, that Heap won 22 out of 32; such a leading majority seems to include some great desiderata, which, from not being present, I cannot elucidate; but if aught can palliate such an unequal contest, it probably may be found in some one or more of the causes here set forth. The two last days certainly were fought uniformly, by winning one and one each day. If Harrison's cocks laboured under no serious injury from their tedious conveyance, whence comes it that a man, whose practice has been considerable, and whose employers, I dare say, are indefatigable, and who are esteemed happy in their choice of cocks, should have suffered such a majority in the three days' fight? It is to me very incomprehensive: these deviations fall to the lot of many practical professors, and perhaps there may be an attendant fate or destiny which Dame Fortune chooses to prescribe, and clog the wheels of men's well-doing. She is a fickle Dame, and is wont now and then to display her vagaries and whims. Who knows but Mr. H. in his next encounter, may be endued with all her gifted dispensations of excelling? Or perhaps she has her limitations, and when professors have run a long course of good and bad—thus far shall ye go, and no farther. Let the young and the active enter the list. Feeders in general have, or exercise, a power to take and refuse such cocks as their employers may think proper to send them; they have their favourite attachments, for one feeder will take in cocks essentially different from another, and you will find some difficulty in accommodating their partialities. It is not every gentleman who makes a match, that is competent to every required qualification to choose a pen of cocks, and he implicitly relies on his feeder; but few embark without experience, and when the particulars are once known, there can be no great difficulty in deciding at once upon such cocks as are adequate to the meeting, and to fulfil the engagement. At any rate, no gentleman can hesitate to take an interesting part, at least jointly with his feeder, and to possess a decided toto, whenever he chooses to pass it. A proper attention to this very circumstance, warrants me to say, it had its agreeable results.

Remarks

On the annexed Pages, columned and prefaced, for the purpose of keeping a regular Account of every Department, with a printed Alphabet.

Page 68 to 67 is to enter every year the brood cocks or stags you have selected for breeding, as well as your hens or pullets, and you will observe to put in the page such cocks, &c. as are taken from, as well as the number they stand in, in their original entry; and as the whole of your stud is at hand previous to your purchasing this book, I have provided several pages under the head of Common Place Memorandums, and of course whatever page they are inserted in will be added to your first year's breeding; and so long as you make use of the same—on the following years you will refer to the page in which they are entered. I have also filled up a form for the guidance of my readers, so that they may be more readily acquainted with the plan laid down for their convenience.

From page 68 to 72 inclusive,
Names and Characters of all my Cocks,

(No. 1 to 5, or as many sorts as you have, with their marks.)

From page 78 to 107 inclusive,
The Persons' Names who walk your cocks, by way of Dr. and Cr.

Page 108 to 114 inclusive,
Bags in Stock.

Page 115 to 124 inclusive,
Common-place Memorandums.

Page 125 to 145 inclusive,
Cash Account.

This account may include all expences, all winnings, &c.—balancing every year.

Pedigree of Brood Cocks, Hens, or Pullets.

THE COCKER.

Pedigree of Brood Cocks, Hens, or Pullets,
1815.

		Turned down at the cottage Black-red Spitfire, page No. 1, No. 1, Cock with four sisters of the cross marks No. —.			
		Hens set.	Eggs.	Produce.	To Hatch.
Mar.	20	Sister	15	18	April 10th.
——	24	Do.	13	11	—— 14th.
April	7	Do.	15	14	—— 28th.
——	14	Do.	11	9	May 4th.
		4	54	47	

		Turned down at Spring Dale Birchin Duck, page No. 1, No. 2, Cock with three sisters, Shropshire Lasses, No. —.			
Jan.	8	Hens set.	Eggs.	Produce.	To Hatch.
Mar.	20	Sister	11	10	April 10th.
——	27	Do.	15	13	—— 17th.
April	1	Do.	13	11	—— 22d.
		3	39	34	

THE COCKER.

Pedigree of Brood Cocks, Hens, or Pullets,

Hens set.	Eggs.	Produce.	To Hatch.

Hens set.	Eggs.	Produce.	To Hatch.

THE COCKER.

Pedigree of Brood Cocks, Hens, or Pullets,

	Hens set.	Eggs.	Produce.	To Hatch.

	Hens set.	Eggs.	Produce.	To Hatch.

Names and Characters of all my Cocks.

NAMES AND CHARACTERS OF ALL MY COCKS.

No. 1. Spitfires   marked.
　 2. Shropshires do.
　 3. Bir. Ducks do.
　 4. Spots   do.
　 5. Piles   do.

NAMES AND CHARACTERS OF ALL MY COCKS.

Account of Cocks, by way of Dr. and Cr.

Dr. CAREFUL, Mr. REPTON,

1814.			No.	Paid	Wlk.
April	12	To bl. red br. norril, stag............	1	—	18
1815.					
May	21	To Birchin duck ot rt. stag	2	—	18
June	10	To br. red, in left, stag...............	8	—	8
	22	To Ginger all fours, stag	4	—	8

Dr. COCKER, Mr. SOUTHWELL,

1814.				Paid	Wlk.

CONTRA, CR.

1815			No.	Won	Lost.
June	6	By black-red to Manchester	1	✗	
Aug.	12	By Birchin to Nottingham	21	—	✗
——	19	By br.-red to Derby	8	✗	
——	28	By Ginger to Litchfield	4	—	✗

CONTRA, Cr.

				Won	Lost.

THE COCKER.

Dr.

		No.	Paid	Wlk.
	Dr.		Paid	Wlk.

CONTRA,				CR.
		No.	Won	Lost.
CONTRA,	Cr.		Won	Lost.

THE COCKER.

Dr.

		No.	Paid	Wlk.
	Dr.		Paid	Wlk.

THE COCKER. 79

CONTRA,				CR.
			No.	Won	Lost.

CONTRA,	Cr.		
		Won	Lost.

THE COCKER.

Dr.

		No.	Paid	Wlk.
Dr.			Paid	Wlk.

CONTRA, CR.

			No.	Won	Lost.

CONTRA,   Cr.

				Won	Lost.

THE COCKER.

Dr.

	No.	Paid	Wlk.
Dr.		Paid	Wlk.

THE COCKER.

CONTRA,			Cr.
	No.	Won	Lost.

CONTRA,		Cr.	
		Won	Lost.

THE COCKER.

Dr.

	No.	Paid	Wlk.

Dr. | | Paid | Wlk.

CONTRA,   Cʀ.

		No.	Won	Lost.

CONTRA,   Cr.

			Won	Lost.

THE COCKER.

Dr.

	No.	Paid	Wlk.

Dr. | Paid | Wlk.

THE COCKER. 87

CONTRA,   Cr.

			No.	Won	Lost.

CONTRA,   Cr.

			Won	Lost.

THE COCKER.

Dr.

		No.	Paid	Wlk.
	Dr.		Paid	Wlk.

THE COCKER.

CONTRA, Cr.

	No.	Won	Lost.

CONTRA, Cr.

		Won	Lost.

THE COCKER.

Dr.

	No.	Paid	Wlk.
Dr.		Paid	Wlk.

THE COCKER.

CONTRA,			Cr.
		No.	Won	Lost.

CONTRA,	Cr.		
		Won	Lost.

THE COCKER.

Dr.

	No.	Paid	Wlk.
Dr.		Paid	Wlk.

CONTRA,				Cr.
			No. Won	Lost.

CONTRA,			Cr.	
			Won	Lost.

94   THE COCKER.

Dr.

		No.	Paid	Wlk.
Dr.			Paid	Wlk.

CONTRA,				Cr.
			No.	Won	Lost.

CONTRA,		Cr.		
			Won	Lost.

THE COCKER.

Dr.

| | | No. | Paid | Wlk. |

Dr.   Paid | Wlk.

THE COCKER.

	CONTRA,		Cr.
		No. Won	Lost.

	CONTRA,	Cr.	
		Won	Lost.

THE COCKER.

Dr.

	No.	Paid	Wlk.

Dr. | Paid | Wlk.

THE COCKER.

CONTRA,			Cr.
	No.	Won	Lost.

CONTRA,		Cr.	
		Won	Lost.

THE COCKER.

Dr.

		No.	Paid	Wlk.
	Dr.		Paid	Wlk.

CONTRA,			Cr.
	No.	Won	Lost.

CONTRA,	Cr.		
		Won	Lost.

THE COCKER.

Dr.

		No.	Paid	Wlk.
	Dr.		Paid	Wlk.

THE COCKER. 103

CONTRA, Cr.

	No.	Won	Lost.

CONTRA, Cr.

		Won	Lost.

THE COCKER.

Dr.

		No.	Paid	Wlk.

Dr. | Paid | Wlk.

CONTRA,   Cr.

			No.	Won	Lost.

CONTRA,   Cr.

				Won	Lost.

THE COCKER.

Dr.

	No.	Paid	Wlk.
Dr.		Paid	Wlk.

CONTRA,			Cr.
	No.	Won	Lost.

CONTRA,	Cr.		
		Won	Lost.

Bags in Stock.

BAGS IN STOCK.

1814.   Bags marked S. S. No. 1 to 70.

	Sent.		*Returned.*	
May 17	To Nottingham	7	May 14th	
June 11	To Leicester	9	June 16th	
—— 20	To Manchester	12	—— 27th	
—— 22	To Loughborough...	9	July 4th	1 kept.
—— 26	To London	11	—— 8th	2 do.
—— 29	To Stamford	22	—— 17th	3 do.
		70		6 kept.

BAGS IN STOCK.

	Bags marked	
	Sent.	*Returned.*

BAGS IN STOCK.

		Bags marked	
		Sent.	*Returned.*

BAGS IN STOCK.

	Bags marked	
	Sent.	*Returned.*

Common-place Memorandums.

COMMON-PLACE MEMORANDUMS.

Brought up from my last Brood Book

    Cocks.

    Hens.

    Pullets.

COMMON-PLACE MEMORANDUMS.

COMMON-PLACE MEMORANDUMS.

COMMON-PLACE MEMORANDUMS.

COMMON-PLACE MEMORANDUMS.

COMMON-PLACE MEMORANDUMS.

COMMON-PLACE MEMORANDUMS.

COMMON-PLACE MEMORANDUMS.

Cash Account.

CASH, Dr.

1815.			£.	s.	d.
		To cash from last balance	872	9	8
June	1	Neated from Manchester	50	1	2
	26	To received for 12 pair to London......	18	0	0
July	6	Neated from Stamford	20	0	0

CASH,   Cr.

1815.			£.	s.	d.
		By paid for walks 1814	15	0	0
June	2	By three brood places	9	2	0
		By my feeder two years	60	0	0
	22	By new bags	2	0	0
	27	By pens new for hospital cocks	4	0	0

CASH, Dʀ.

		£.	s.	d.

THE COCKER.

CASH, Cr.

	£.	s.	d.

CASH, Dr.

			£.	s.	d.

CASH, Cr.

			£.	s.	d.

CASH, Dr.

		£.	s.	d.

CASH, Cr.

			£.	s.	d.

CASH, Dr.

		£	s.	d.

CASH, Cr.

£. s. d.

CASH, Dr.

			£.	s.	d.

CASH, Cr.

		£.	s.	d.

CASH, Dʀ.

		£.	s.	d.

THE COCKER.

CASH, Cr.

			£.	s.	d.

CASH, Dr.

		£.	s.	d.

CASH, Cr.

		£.	s.	d.

CASH, Dr.

		£.	s.	d.

CASH, Cr.

		£.	s.	d.

CASH, Dr.

		£.	s.	d.

CASH, Cr.

	£.	s.	d.

Alphabetical List.

ALPHABETICAL LIST.

A	**C**
B	**D**

ALPHABETICAL LIST.

E	G
F	H

ALPHABETICAL LIST.

I	L
K	M

ALPHABETICAL LIST.

N	P
O	Q

THE COCKER. 153

ALPHABETICAL LIST.

R	T
S	V

ALPHABETICAL LIST.

W	Y
X	Z

www.ingramcontent.com/pod-product-compliance
Lightning Source LLC
Chambersburg PA
CBHW062215220526
45471CB00009B/3216